LIGHT RAILWAY CONSTRUCTION

By
E. R. CALTHROP ESQ.,
C.E., LIVERPOOL.

PLATEWAY PRESS

LIGHT RAILWAY CONSTRUCTION

ISBN 1 871980 33 X

© Plateway Press 1997

All rights reserved.
No part of this publication may be reproduced,
stored in a retrieval system, or transmitted,
in any form or by any means, electronic,
mechanical, photocopying, recording or otherwise,
without the prior permission of the Publisher.

Printed by Postprint
Taverner House, Harling Road, East Harling, Norwich, NR16 2QR

PLATEWAY PRESS
Taverner House, Harling Road, East Harling, Norwich, NR16 2QR

CONTENTS

Acknowledgements iv

Introduction v

Theory of Narrow Gauge Light Railways 1

Demonstration of the Correctness of the Theory 4

History of the Barsi Light Railway 8

Traffic on the Barsi Light Railway 10

The Barsi Permanent Way 11

The Barsi Rolling Stock – Locomotives 15

The Barsi Rolling Stock – Wagons 21

The Barsi Rolling Stock – Carriages 26

Means for avoiding Transshipment at Junctions
Between Broad and Narrow Gauge Railways 28

The Barsi Rolling Stock as Applicable
to Military Operations 29

Addendum 32

Acknowledgements

In preparing this reprint, we are greatly indebted to
Mr J. H. Price for the loan of the original,
and to Messrs. I. Fraser, R. Humm, R. N. Redman and K. Taylorson
for help with illustrations and additional material.

INTRODUCTION

During the latter part of the 19th century, the narrow gauge light railway was seen as an important tool in the economic development of rural areas in both European countries and their overseas colonies.

Like his contemporary, Sir Arthur Heywood, Everard Calthrop sought to design such railways for maximum efficiency from first principles, rather than as either miniature main line railways or lightly laid rural tramways.

This paper, written in 1897, deals primarily with the Barsi Light Railway in India, opened in stages between March 1897 and December 1906, which became the showpiece for the extensive network of 2ft. 6in. gauge railways built throughout the Indian sub-continent. At Barsi, Calthrop demonstrated his theories – namely that 2ft. 6in. was the ideal gauge for a maximum tonnage to be carried at minimum cost, utilising a common axle loading of five tons for both locomotives and rollingstock. A rail weight of 30 lbs. per yard on a well built trackbed allowed efficient working of traffic for a modest capital cost.

Calthrop's name will always be associated with the Leek & Manifold Valley Light Railway in Staffordshire, opened in 1904. Even though that line was an economic failure it vindicated his theories for the track needed no major renewals throughout the line's 30 year life. Here Calthrop introduced transporter wagons to eliminate the delays and costs of transshipment at the break of gauge. Sadly, no other British line adopted this idea, although its widespread use in Europe has kept many lines open which otherwise would have been closed or converted to standard gauge many years ago.

Calthrop later acted as consulting engineer to other similar lines in India and Barbados. As mentioned in the paper, he designed and patented a special arrangement of swivelling head for the Jones centre buffer coupler. Outside his railway interests, he was a noted authority on the breeding of Arab horses and patented the 'Guardian Angel' parachute that saved many fliers' lives on the Allied side during World War 1. He died on March 29th, 1927.

October 1997 *Andrew Neale*

A train on the Barsi Light Railway, 2ft. 6in. gauge, constructed in India to Calthrop's design. The flat wagons are carrying a Galloway boiler 28ft. 6in. long x 7ft. 6in. diameter, and weighing 14 tons, plus other heavy machinery. It should be noted that the G.I.P. system, with which the Barsi connected, required two standard gauge wagons to carry the boiler load.

THE TRAFFIC TO BE CARRIED in any district, through which a railway is about to be constructed, is the same whether the proposed railway be built on the standard gauge, or on a very narrow gauge; but the revenue, which might result in an actual loss on the working of the former, may be made to produce a satisfactory return on the much smaller capital of the latter. The object aimed at by the advocates of light railways of very narrow gauge is the provision of railway communication on a remunerative basis, in countries and districts where the traffic to be carried is considerable, but insufficient to produce a proper return on the capital required for a railway of normal type.

To attain this object the fundamental idea of light railway construction and working is the elimination of every kind of expenditure which is non-essential to its efficiency as a means of transport, and the reduction of all permanent way, plant, and appliances to their simplest and most inexpensive forms.

The reductions in cost of permanent way, obtained by the use of lighter rails and smaller sleepers, are supposed by many to be the only advantages offered; but while the gains under these heads are not inconsiderable they are really the least of all. The principal savings in capital cost, effected by narrowness of gauge, occur through the much greater flexibility of its alignment as compared with that of the wider normal gauges. The established minimum radius of curves, on open line and in sidings or difficult country, is for the several gauges respectively as follows:-

Minimum Radius of Curve (Indian Practice)

	Gauge	Radius in ordinary country	Radius in sidings or difficult country
	Ft. In.	Feet	Feet
Standard	5 6	1600	600
Metre	3 3⅜	1000	400
*Special	{ 2 6	250	150
	{ 2 0	150	60

Minimum Radius of Curve (English Practice)

Normal	4 8½	1320 = 20 chains	462 to 660 = 7 to 10 chains

The small radius of the curves of the narrowest gauges confers immense advantages in locating the alignment of a light railway. Its flexibility permits it to wind in and out to avoid deep cuttings, heavy embankments, and the severance of valuable property, and, if required, to follow all the convolutions of an ordinary road. Tunnelling, heavy rock cuttings, and steep gradients can often be avoided by a quick detour quite impossible on the standard gauge; the latter by the greater stiffness of its alignment is often compelled to plough through buildings and other valuable property, and to construct bridges and other expensive works, which otherwise might be avoided. The extraordinary cheapness in capital cost of the 2ft. 6ins. gauge lines built in India is explained chiefly by this avoidance of costly obstacles, and the utilisation of long reaches of ordinary roads which Government has permitted them to use free of cost.

---

* In the new regulations of the Government of India relating to standard dimensions to be observed on 2ft. 6in. and 2ft. gauge railways in India, the minimum radius permitted on main lines is 238ft. = 24° angle of curvature, and 204ft. = 28° angle of curvature respectively, but in open country a radius of 1,432ft. is recommended for both.

Compared with the *possibilities* of standard gauge rolling stock, the cost of carriage stock per passenger seated, and of wagon stock per ton of load, is somewhat less. As compared with the standard gauge wagon stock *actually in existence*, the cost per ton of load is very considerably less. The cost of locomotives per ton weight and per indicated horsepower is, as one would expect, rather greater on the narrow gauge.

In designing the permanent way and rolling stock for a narrow gauge light railway, the principal object to be aimed at is to obtain the greatest proportionate traffic capacity, from the cheapest possible track. Everything

Above: Two Barsi Light Railway locomotives. On the left TANSA, a Hunslet 0-6-0T, makers number 430 of 1887. Supplied to Walsh Lovett & Co., India, the loco was purchased by the Barsi and used during the construction of the line. The loco on the right, 0-8-4T Sir R. E. Webster, Kitson 3649 of 1896 was the second of four identical locos (Barsi class A) supplied for the opening of the line in 1897. A fifth was supplied in 1898.

Below: Thereafter, the design was modified to a 4-8-4T (Barsi class B - SIR ALEC, Kitson 4328 of 1905), twelve in all being built. Both class A & B were classic Calthrop designs restricted to a maximum loading of five tons per axle.

must be made subsidiary to the cheapening of the track. Any additional costliness of the rolling stock, as will be shown later, is as nothing if it results in any considerable reduction of the weight, bulk and cost of the track.

On any kind of railway, the cost of track must be the principal consideration, for it is obvious that, if it is possible to quadruple the length of the track for the same expenditure of money, by so doing the traffic area of the line is also quadrupled. In other words, a narrow gauge light railway, laid down at one-fourth the cost of the standard gauge per mile, is sixteen times more efficient as a revenue producer, than a standard gauge line costing the same amount, by means of its being four times as long and carrying four times the traffic over the additional length. I am speaking, of course, of conditions where the traffic is an even quantity and proportional throughout to the length of the line, and where the cost per mile is fairly even throughout its length. Such conditions, from a variety of causes, are comparatively rare in England, but in other countries, possessing large areas unoccupied by railways, they are by no means unfrequent.

It follows then, under these conditions, that a very narrow gauge line being able, with the same expenditure of capital on track, to earn sixteen times the revenue of a standard gauge line, it can afford, theoretically, to run through country so poor that it produces but one sixteenth of the traffic that would be necessary to obtain a remunerative return on the capital of a standard gauge line if constructed through the same country.

After what has been said it is hardly necessary to point out that the prospects of any light railway improve directly with its length. The longer the run the greater the traffic and the cheaper it can be worked. If a light railway, constructed at a minimum cost per mile, should not make an altogether satisfactory return on capital, the surest remedy is to extend it.

Demonstration of the correctness of the theory

How is this theoretical superiority of narrow gauge lines as dividend earners borne out in practice?

In India as is well known there are now four gauges: The Standard 5ft. 6in.; the Metre 3ft. 3⅜in.; and two Special gauges, the 2ft. 6in. and 2ft. 0in. In these last days of the century with the results of 60 years of railway progress behind us, it is difficult to understand the failure 50 years ago of the Government of India to appreciate the certainty of the linking up of the

The original carriage stock, Barsi Light Railway. These coaches were 40ft. long over the headstocks, 43ft. 3in. over the centre buffers and 7ft. 6in. wide overall. The 3rd class cars had eight compartments with Indian Govt. regulation seating capacity of 64 passengers. There were also composite coaches of the same general dimensions, consisting of 1st and 2nd class compartments, a guard's compartment and P.O. accommodation. The 1st class was provided with sleeping arrangements. Both 1st and 2nd class compartments had lavatories adjoining.

European and Indian Railway systems, which is now on the threshold of events. The original gauge, unfortunately selected as the standard for India, was the 5ft. 6in. At a latter date it was discovered that the costliness of the 5ft. 6in. gauge was an absolute bar to the extension of railway communication in certain of the poorer districts of India, and after much agitation and consideration the metre, or 3ft. 3⅜in. gauge, was eventually sanctioned for lines, which were held to be of minor importance strategically, with results which generally fully bore out the anticipations of their supporters.

More recently again attempts were made to obtain a still cheaper form of railway in districts where even the metre gauge could not be laid down remuneratively, with the result that after a still greater resistance on the part of the authorities a number of short lines have been sanctioned and constructed on the 2ft. 6in. gauge.

LIGHT RAILWAY CONSTRUCTION

There is perhaps no country in the world where railway statistics are as carefully registered and collated as in India, and there is certainly no more interesting production published by any Government than the Administration Report on the Railways in India, which is characterized by intelligent analysis and extreme accuracy.

In this Report statistics of the working of the standard, metre, and special gauges are dealt with separately, so that I have been able to represent the general results of the working of all railways in India grouped according to their gauges. The subjoined table, in which the ultimate influence of gauge upon capital cost, working expenses, and net profits is shown to be of a most remarkable character, affords the completest justification of the policy of the Secretary of State in anticipating the recent legislation in England by permitting the construction of railways of a smaller gauge than the normal, in districts where the traffic is insufficient to support a railway of standard type.

General results of the working of all railways in India, for the year ending December 31, 1895.

Particulars	Standard Gauge 5ft. 6in.	Metre Gauge 3ft. 3⅜in.	Special Gauges 2ft. 6in. & 2ft. 0in.
Capital Cost			
Average cost per mile open... Rupees	158,730	71,121	32,950
Ditto at Exchange 1/3¼	(£10,095)	(£4,519)	(£2,093)
Passengers			
Passenger unit miles, per mean mile worked ... Unit-Miles	368,456	264,351	93,943
Average distance of journey ... Miles	41.9	37.6	27.1
Goods			
Goods, ton miles, per mean mile worked...Ton-Miles	358,028	124,511	16,364
Average distance carried ... Miles	156.1	119.2	34.1
Working Expenses			
Percentage of working expenses on gross earnings, per cent.	45.14	49.12	53.16
Net Profits			
Percentage of net profits on total capital outlay on open line, per cent.	5.78	5.73	7.67

The facts disclosed by the figures as they relate to the three gauges are as significant as they are unimpeachable. As compared with the average cost of the standard gauge, it is shown that the average cost of a metre gauge line is rather less than one half, while the cost of a 2ft. 6in. gauge line is actually only one-fifth. It should be also noted that with a difference of only 9⅜ins., the metre gauge is more than twice as costly as the 2ft. 6in., although in ultimate traffic capacity the difference between these two gauges is extremely small. These figures include rolling stock and all capital outlay whatever. But taking into consideration the circumstance that when the older standard gauge lines were built there was a higher range of prices for rails and rolling stock, we are nevertheless face to face with the fact that for the same expenditure of money it is possible to construct and equip 400 miles of 2ft. 6in. gauge at the cost of 100 miles of standard gauge. In India at all events it is not open to doubt that a line of narrow gauge, carried to a distance of 100 miles, must produce, in its combined capacity as a feeder to the main trunk line, and as a distributor of English merchandise, a much greater effect upon the development of an unopened district than a standard gauge branch of 25 miles, with 75 miles of cartage behind it. The extra length of the line secures for the benefit of the country much cheaper carriage, and for itself a much greater traffic catchment area. The cost of transshipment at the junction with the main line, which however can be altogether avoided by an appliance about to be described, is as nothing compared with the saving effected in the cost of carriage over the additional 75 miles of line which can be laid for the same money.

The next and most important result which the tables makes clear is that the 2ft. 6in. gauge running through the poorest districts, for short distances only, with a passenger traffic per mile of less than *one-fourth* that of the standard gauge, and with a goods traffic per mile only *one-twentieth* that of the standard gauge, is not merely able to survive, but can actually show a greater percentage of net profits on total capital outlay than the standard gauge trunk lines running through the pick of the country, and backed by all their great volume of arterial traffic carried over long distances. The above results illustrate in the most practical manner the concrete effect of the multitude of small advantages gained by the adoption of a very narrow gauge, and fully demonstrate the correctness of the claim made for it that such a line is able to run with satisfactory financial results through country producing but a very small fraction of the traffic required to obtain a reasonable return on the capital of a standard gauge line.

It is as well to point out that there is a great principle underlying the question of gauge: *A railway is a machine, and, like any other machine, is economical only when working within a reasonable measure of its full power.* In the recognition and observance of this principle lies the whole art and mystery of the financial success which has attended the working of narrow gauge lines on the Continent and in India, in districts where a standard gauge line would not only starve, but would lose money to the end of the chapter.

History of the Barsi Light Railway

The main points in the history of the Barsi Light Railway may be of interest to you as exemplifying the difficulties which have attended the introduction of Light Railways in India, and of this project in particular.

The Barsi route, as the main line of traffic the between dominions of His Highness the Nizam of Hyderabad and the Port of Bombay, has existed from time immemorial. Previous to the construction of the Great Indian Peninsula Railway the traffic was carried down to Bombay in bullock carts and on pack animals. In the year 1856 the existing rough cart track was converted into an unbridged and roughly ballasted road with the result of a great increase in traffic. After the construction of the South East branch of the Great Indian Peninsula Railway in 1860, the traffic increased further to an average of 500 carts daily. In 1862 proposals were submitted to the Government of Bombay by the Chief Engineer of the Bombay Presidency for the construction of a bridged and embanked road to be traversed by a light railway. In 1870 the road and bridges were constructed, but the Government of Bombay did not find themselves in a position to incur the cost of the light railway. The improved condition of the road was followed, however, by a further great increase in traffic. In 1878 the Government of Bombay proposed to the G.I.P. Railway that a branch line in Barsi should be constructed as an extension of their system, and surveys were made in 1880 and 1881; but after all preparations had been made the Secretary of State finally informed the Company that he was "not disposed to enlarge the operations of the Company under any arrangement which would involve any extension of the guarantee of interest, or of the grant of pecuniary assistance

Interior of Saloon Car, Barsi Light Railway. A special saloon car was provided for the use of Government officials and other distinguished passengers. It comprised an inspection compartment, a long dining and sleeping saloon, an entrance and luggage vestibule, a lavatory and bathroom with hot and cold water, a kitchen and a servants' compartment.

to the Company from the State in any shape." The project of the branch as an integral portion of the G.I.P. system was therefore abandoned. A subsequent proposal that the line should be constructed by Government as a provincial railway, and worked by the G.I.P. Railway was introduced in 1882 by the Government of Bombay, and abandoned in 1885 with a notification that since Government found itself unable to advance the capital required the line should be undertaken by private enterprise.

In 1887, being Assistant Locomotive Superintendent on the G.I.P. Railway, I undertook a preliminary survey, and having satisfied myself as to the prospects of traffic, applied to the Government of Bombay for the concession to construct a line along the existing road from Barsi Road

Station on the G.I.P. Railway to the town of Barsi, a distance of 22 miles, on the 2ft. 6in. gauge, without any guarantee of interest. The negotiations with Government were continued without intermission from March, 1887, until July, 1895 – a period of more than 8 years – during which I arranged three times for the capital required and expended £6,300 in surveys, estimates, and two special visits to India, involving together an absence from England of two years and ten months. It has been an experience which I do not wish to repeat, and although no guarantee or financial assistance whatever had been asked for, it cost over £260 per mile of line merely to obtain the concession on these terms.

In looking back on these protracted proceedings my pleasantest recollections are of the help extended to me by the Consulting Engineers to Government, and of others who were in entire sympathy with the system which I advocated. Without their energetic assistance it would have been impossible to have overcome the influential obstruction which then existed. It is pleasing, however, to be able to state that the resistances have now disappeared, and that the much talked of "encouragement of private enterprise," which for many years had been a mere snare and a delusion, is now becoming an accomplished fact.

Traffic on the Barsi Light Railway

Before proceeding to describe the nature of the permanent way and rolling stock which I have designed for the Barsi Light Railway, it is desirable to give you an impression of the amount and character of the traffic to be accommodated. We are putting a small feeder line into a tract of country as large as the whole of England and Wales, which up to the present is totally unprovided with railway communication of any kind. The town of Barsi is the central mart of the trade to and from the great expanse of territory in the valley of the Godaveri belonging to His Highness the Nizam of Hyderabad, which territory is cut off from any access to the G.I.P. Railway for nearly 100 miles by the rivers Bhogavati and Sina, except through Barsi. Complete records of the traffic received at the Barsi Road Station have been kept for many years, as also of the traffic passing through Barsi town. For the five years ending December, 1894, the average goods traffic at Barsi Road Station passing over the G.I.P. Railway amounted to 77,599 tons per annum, of which 95 per cent. was estimated to pass through Barsi town, an estimate which agrees with the municipal records. Since the year 1881, the first of which I possess complete records, the largest traffic occurred in 1891, the amount being 107,000 tons, and the smallest in 1888, the amount being 51,000 tons. The import traffic from Bombay averages one-third, and the export traffic from Barsi two-thirds, of the total tonnage. The extent of the area served by Barsi as a centre of distribution will be better realised from the fact that the average weight of salt imported, for human consumption alone, during the above mentioned five years, exceeded 7,000 tons annually. Other imports than salt consist chiefly of iron and other metals, machinery, mineral oil, silk and cotton piece goods, and twist and other European products. The principal article of export is cotton, of which the output in several seasons has exceeded 90,000 bales of cotton of $3\frac{1}{2}$ cwt. each, pressed at Barsi. The remaining exports are wheat, linseed, ground-nuts, native food, grains of all kinds, gingelly and other oil seeds, indigo, dyes, sugars, spices, wool, hides and skins. The passenger traffic, including pilgrims, over the road traversed by the light railway amounts to about 200,000 each way per annum.

The means of transport between Barsi and the G.I.P. Railway has been entirely confined to bullock carts carrying from 6 to 12 cwt. each, according to the nature of the goods. In the busy season 1,000 to 1,500 carts leave Barsi daily for the station. Beyond Barsi, bullock carts are employed where the roads are fairly good, but pack animals carrying from 1 to $2\frac{1}{2}$ cwt. each are still used to some considerable extent.

The Barsi Permanent Way

In the selection of the most suitable gauge for the Barsi Light Railway many considerations had place. It was essential, in view of the very large existing traffic and of its subsequent expansion, that its ultimate traffic capacity should be very great. On the other hand it was equally necessary that the cost per mile of permanent way should be so small that it would pay us to extend the line into localities where the traffic was comparatively insignificant, and that we should be in a position to offer special siding accommodation with considerable generosity. It is accommodation and convenience which attract traffic nowadays, and the ability to run sidings at a comparatively trifling cost, into the cotton presses, warehouses, godowns, and compounds so as to eliminate all avoidable handling and cartage, all tends to develop trade and increase the traffic over the main line itself. These and the considerations enumerated at the beginning of this paper, originally led me to select the 2ft. 6in. gauge as the best for the purpose, and the more I know of it the better I like it. There is no doubt that as compared with all others it is the gauge possessing the *greatest carrying capacity per cent of cost of track*. It has sufficient stability to carry goods of very great weight and bulk, while the flexibility of its alignment is such that it can accommodate itself to country of the most mountainous and difficult character, at a fraction of the cost of a standard gauge line negotiating similar difficulties.

After selection of gauge the next point was the minimum weight of rail and maximum load to be placed upon it. This, with the concurrence of the Secretary of State for India, I fixed at 30lbs. per yard and 2½ tons respectively giving a maximum load of 5 tons per axle.* On the length to Barsi town the weight of rail actually laid down is 35lbs, but this has been purely a matter of economy in maintenance under an extra heavy traffic. For branches or sidings where the traffic is comparatively light we have utilised the 30lb. rail only.

In the 30 and 35lb. Barsi sections the head of the rail is of the same width, the extra metal in the latter being put principally on the top of the head. Fish plates and fish bolts are identical for both sections of rail. The head of both rails is made particularly wide with the object of providing ample bearing surface, and preventing grooves being worn in the tyres. Most sections of light rails err in this respect, with the consequence that they cut the tyres badly. The sleepers are of stamped steel weighing 40lbs. each. They have been spaced at distances so arranged that there is equal resilience both as regards rail joints and throughout the intermediate length of rail. This practice was first determined by theoretical considerations, the correctness of which was afterwards demonstrated by a series of lever experiments, conducted on the permanent way by means of a lever testing machine, and

---

* In the new regulations of the Government of India, issued since the construction of the Barsi Light Railway, for 2ft. 6in. and 2ft. gauge railways, the maximum axle-load on 30lb. rails has been fixed at 6 tons for locomotives, and 4 tons for goods and coaching stock. Mechanically, this is, of course, an absurdity, but the unnecessary limitation of axle-load on goods and coaching stock has probably been conceived with the object of discouraging the construction of railways of gauges narrower than the metre, since it debars them of making full use of their natural traffic capacity.

finally by practice. At the Newlay Exhibition the smoothness of the track, laid on this principle, was generally remarked. The points and crossings are 325 feet radius, the angle of crossing being 1 in 8, and can be laid down either as right hand or left hand turnouts. The intermediate rails are bent, cut to sizes and fitted, and are shipped complete with each set, so that, on arrival, they can be laid exactly as they are.

Before leaving the subject of permanent way, I wish to emphasise the necessity, in regard to light railway work, of abandoning the extremely high standard of construction considered necessary on standard gauge railways, and adopting a standard which, while it takes nothing away from the efficiency of the line as a means of transport, reduces the cost per mile to a very appreciable extent.

One of the advantages of the 2ft. 6in. gauge is that it requires no raised platforms. All that is necessary is a well drained gravelled space, edged with stone slabbing to mark its boundaries. Compare this with the cost of paved platforms raised two or three feet above rail level for a length of 600 to 800 feet, to which is added the cost of the extra height of the foundations of all station buildings placed upon such platforms. On the standard gauge, and with English types of carriage stock, such platforms are practically a necessity on account of the great height of carriage and wagon floors above rail level. For light railway work, I am against all ornamental and unnecessary expenditure, particularly as regards the erection of permanent buildings, for the accommodation of employees, on a scale of extravagance altogether above the kind of habitations in which they are accustomed to live, and also as regards lavish accommodation at stations before traffic requirements are thoroughly tested.

In my opinion the fact that it is essential to their financial success that narrow gauge light railways have to be built and worked on principles and guided by rules totally divergent from those in use on standard gauge railways is a very strong argument in favour of their independent administration. The whole bent and training of the rank and file of the staff of a standard gauge line is towards solidity and lavish expenditure, and with the advent of heavier train loads and higher speed this tendency will become more and more pronounced. The metier of the light railway man is, on the other hand, to eliminate expenses, superfluity, and complexity in every shape and form, and to evolve a type of line on which efficiency of action is combined in every department with the greatest simplicity of equipment. To place the working of a narrow gauge line, on which such a policy is required, in the hands of a staff furnished, changed and controlled by an adjoining standard gauge railway, and to expect it to carry out a rôle totally opposed to all its previous traditions is not likely to produce the desired results, and, so far as I am aware, has never yet done so.

As it is practically impossible to effect any useful reductions in the cost of standard gauge permanent way, I am satisfied that if any considerable development of the light railway system is to take place in England, it must be on the narrow gauge, so as to obtain the full maximum advantage as regards first cost. In this event it is of the very greatest importance that one standard gauge should be fixed for such narrow gauge lines, so that rolling stock may be interchangeable. It is inevitable that light railways, although beginning with isolated projects, will join up to one another. I go further than this and say that light railways in England will be as unsuccessful financially as those in Ireland, unless some central organisation can be brought into being to undertake their management and working. The inordinate management charges following on the independent direction of undertakings of extremely short mileage, together with the abnormally heavy capital cost per mile of the Irish 3ft. gauge light railways, are responsible for the generally unsatisfactory character of returns on capital invested. Greatly improved results would follow were all these Irish light railways to be amalgamated into one large undertaking, so that the proportion of management charges to other expenses could be reduced, together with the cost of the maintenance and renewal of rolling stock. At present every line has its own little workshops, with its own patterns of rolling stock, several patterns existing sometimes on one line. When new engines or wagons are required they are bought one or two at a time on very disadvantageous terms of payment. All this should be swept away. Under one large company, or central management, the rolling stock should be standardised, so as to interchange on every line of the same gauge, and with larger quantities of engines, carriages, and wagons of standard types, they could be purchased and repaired with much greater economy than obtains at the present time. Were this to be done the financial standing of Irish light railways would assume a very different aspect. As it is, these small lines, working under the above named disadvantages, are nevertheless doing extremely good work in developing the resources of districts beyond the reach of ordinary railway communication, and, notwithstanding the baronial rates that are levied in most cases to supply the deficiency between profits and the guaranteed interest on capital, it is certain that they are popular.

The Barsi Rolling Stock - Locomotives

The principal novelty in the Barsi Rolling Stock has been the adoption of a uniform working axle-load throughout for engines, wagons, and carriages; the working axle-load being also the maximum adopted for the rail, namely five tons. On all other light railways, and indeed I may say on all railways with which I am acquainted, it is customary for the engine axle-loads to be much greater than those in use on the carriage and wagon stock.* On the 3ft. gauge light railway in Ireland, for instance, of which there are now eleven with a total length of 2,025 miles, the maxima axle-loads on carriage and wagon stock vary from three to four tons, while the engines have axle-loads of eight and nine tons, requiring rails of 50lbs. per yard, or exactly double what would have been necessary if the engine axle-loads had been maintained at a figure approximating to the maximum axle-load of the carriages and wagons. The additional weight on the wheels of these Irish engines was totally unnecessary, as has been proved by the Barsi engines, whose total adhesion and tractive force is greater, notwithstanding their much smaller axle-load. To get the greatest cubic and load capacity out of narrow gauge wagon stock, the maximum axle-load must be utilised to its full extent. We arrive therefore at a principle in light railway construction which is that the greatest economy, *i.e.*, the maximum carrying capacity on the minimum weight of rail can be secured only by uniformity of axle-load.

The next point to which I would direct attention is that the whole of the Barsi rolling stock has been constructed on a system of standard details, so that like articles are interchangeable throughout. Each part of the engine has been most carefully designed, not only with reference to its work but also with regard to its position and the effect of its weight in securing that the maximum axle-load should not be exceeded on any single pair of wheels. Every single detail has been allotted a standard number, which is stamped

---

* See footnote to page 294 ante, on axle-loads.

upon it, so that in the case of damage from accident or in the course of ordinary wear and tear, all damaged parts can be obtained from the headquarters stores depôt, or ordered direct from England merely by cabling their standard numbers. The same principle of the standardisation of parts has been carried out throughout the carriage and wagon stock. This system is productive of extreme economy, not only in construction, on account of the lessened cost of the manufacture of numbers of identical articles, but also in maintenance, as it allows of the behaviour of parts being kept constantly under observation. In the working of rolling stock, designed on these principles, the quantity of any one standard part worn out or repaired each half-year arrives at a figure dependent on the number in use and the train-mileage run. If the number be found to be abnormally high, attention is directed to the subject, and the defect whether of design or material is rectified. It will be seen that when once the relationship of repairs and renewals to train-mileage is ascertained, the quantities of all standard duplicates, required for renewal in any half-year or on any given train-mileage, can be estimated for in advance with very great accuracy. In the case of light railways in India and the Colonies, a line has to draw on England as its base of supplies. Consequently it is most important that the stores can be indented for sufficiently in advance to prevent rolling stock from being incapacitated and withdrawn from traffic because they have not arrived in time. The Barsi rolling stock has been specially designed with reference to its adaptability for military purposes, and I need hardly say how extremely important in military operations it is that repairs should be effected with the least possible delay. Duplicates of all parts particularly subject to wear should be sent out with the first shipment of material. The standard detail system permits of the system of repair by exchange. Exchange is a matter of minutes, while repair is a matter of days and weeks. We arrive now at another principle in light railway construction, namely, that the rolling stock should be of the fewest types, and that each of those types should be built up of the greatest number of standard parts.

The Barsi engine, a view of which is given, is 29ft. 6⅛ins. long over buffers, and weighs 29 tons 8 cwt. in working order and fully loaded. It is of the consolidation type, having 8 wheels coupled of 2ft. 6ins. diameter, and is provided with a four-wheeled swing link bogie truck at the trailing end. The cylinders are 13ins. diameter by 18ins. stroke. The working pressure is

150lbs. The slide valves are balanced and driven by Walschoert's gear, dispensing with eccentrics. The tanks have a capacity of 800 gallons and the coal bunker of four tons. The tractive force of the engine is 11,088lbs. with a cut off at 75 per cent. of the stroke, 10,650lbs. at 70 per cent., and 9,610lbs. at 60 per cent. The slidebars are boxed in to keep out dust, the engine being built for use on ordinary roads. All the wheels, including those of the bogie, are fitted with brakeblocks, operated by hand or steam at will. The engine is fitted with a whistle for train signalling, a siren to warn cattle off the road,

Barsi Light Railway locomotive 'HAMILTON'.

and with a steam bell for use when moving through the streets. The total wheelbase is 18ft. 6ins., but the rigid wheelbase is only 8ft. 3ins., which enables the engine to run round extremely sharp curves. The engine and all other vehicles are fitted with the Jones-Calthrop Patent Flexible Buffer-Coupling. At the Newlay Exhibition this engine took a train load of 180 tons up a gradient of 1 in 57, occurring on reverse curves averaging 200ft. radius. The curves as laid down were parabolic, and at the sharpest point were

equal to a radius of only 175ft. During the trials the weather was generally very bad and the rails greasy, so that the performance, which was repeated many times daily, was a severe test of the powers of the engine. On several occasions the train was stopped on the reverse curves when mounting the gradient, and was re-started without difficulty. The following views show a portion of the experimental train standing on the reverse curves:-
On a level straight line, this engine will draw 1,036 tons at 15 miles per hour.

Above: Barsi train on curve.

Below: Barsi train on S curve.

On the heaviest gradient and curve on the Barsi main line, namely 1 in 100 on 600ft. curve, the engine will haul 291 tons at 8 miles per hour. A train of 280 tons, composed of 13 low-side wagons and 1 brake, fully loaded, will have a tare weight of 65 tons 7 cwts., and will carry 210 tons of goods. As 16 trains a day can be run in each direction without difficulty, the theoretical capacity of a single line with crossing stations is no less than 3,360 tons each way, which is several times what is required at Barsi even in the height of the busy season.

Jones-Calthrop Patent Flexible Buffer. Side view.

The engine is fitted with a central buffer which also forms a coupling. This buffer coupling is a modified form of the Jones Patent Flexible Buffer, adopted as the standard for metre gauge lines by the Government of India, to which I have added rocking dishes which permit it to move laterally through an angle of 36 degrees. The two buffer heads are coupled by means of a central hook, and are drawn together by a right and left hand screw coupling, so that the buffer spindles become practically one solid bar. There is therefore no movement or slack whatever between the two heads. When a train is going ahead the two outer springs are in compression; in shunting

Jones-Calthrop Patent Flexible Buffer. Top view.

the two inner springs are compressed. The buffer spindles are not at any time in rubbing contact with any part of the headstock, being supported only at the smaller ends of the springs, and, as there is also no rubbing of the heads, the wear and tear of the buffer is practically nil. It will be seen that this Jones-Calthrop buffer is what is technically termed a tight coupling, and, as there is no slack between the heads, it is impossible to cause damage by snatching, which is an important feature, since my experience of wagon repairs shows that 90 per cent. of the withdrawals of wagon from traffic are due to damages to draw-gear caused by snatching slack couplings when starting a train, or in taking up the slack at the bottom of inclines. The large angle through which the rocking dishes permit the buffer to move, has enabled me to put cars 40ft. long on this narrow gauge. The principal advantage of the buffer, however, is derived by the great freedom of the lateral movement, by means of which flange friction round sharp curves has been reduced to such an extent as to very appreciably increase the train load. The illustration opposite is a top view of a buffer between two 40ft. coaches standing on a curve of 175ft. radius, which shows the angles made by the ends of the coaches to each other and to the track, and also the angle of the buffer to the headstocks.

I have no doubt that the large excess which our engine at Newlay drew over its theoretical load, calculated on the usual formulae, was due in part to the reduction, as explained above, of resistances on sharp curves.

The Barsi Rolling Stock Wagons

I now come to a description of the tare weights of the Barsi wagon stock, which have been designed to combine the greatest carrying capacity with the minimum tare weight, and which, together with the great power of the engines, formed the chief point of attraction at the Newlay Exhibition. It has been explained that in designing the permanent way and rolling stock for a narrow gauge light railway the principle object to be aimed at is to obtain the greatest proportionate traffic capacity from the cheapest possible track. With a given engine power, any reduction in the dead weight of the train produces a corresponding increase in its carrying capacity, and our object at Barsi with our large traffic has been to carry the heaviest goods load per train on the lightest tare weight compatible with a proper reserve of strength, and a due regard to the cost of maintenance and repairs. After careful examination of various systems claiming to effect reductions in the tare weights of rolling stock, I ultimately selected that known as Fox's Pressed Steel Underframes. By the adoption of pressed steel I have been able, while working to the uniform maximum axle-load of five tons, to design open wagons, 25ft. long over headstocks, with a paying load of 15 tons 18 cwt. and a tare weight of only 4 tons 2 cwt., the percentage of tare to the total being only 20.5 per cent. This low side wagon was placed under a test load of 40 tons of pig iron, with a temporary deflection of $5/16$ths of an inch between bogie centres, and without a trace of any permanent set. A second wagon was placed under the same load of pig iron at the Newlay Exhibition with the same deflection, where it remained for a week, and when unloaded there was again no trace of permanent set. The high side open wagons weigh 5 ton 7 cwt., and carry 14 tons 18 cwt. of goods. The covered wagons weigh 5 tons 18 cwt. and carry 14 tons 2 cwt. While I do not deny that other

Low side wagon, 25ft. long x 7ft. wide, tare weight 4 tons 2 cwt., loaded with test load of 40 tons of pig iron. Deflection in centre, $^5/_{16}$in. permanent set, nil.

systems have their merits, I know of none other which can produce such results as these while maintaining simplicity of detail and the same reserve of strength.

Light tare weight effects a permanent economy in working expenses by making it possible to carry a greater quantity of goods per train at the same cost of coal, oil, and wage, and by reducing the wear and tear on permanent way. This is an actual saving of revenue, day by day and year by year, in respect of every train that is run. In regard to capital cost the results are of the highest possible importance. Had our wagon stock been built with the usual heavy tare weights it would have been necessary, in order to carry the same weight of goods in each train, to have adopted a greater axle-load, to have designed much heavier and more powerful engines, and to have employed rails and permanent way of much greater weight per yard.

To secure these advantages it would be worthwhile, if it were necessary, to pay more per vehicle for wagon stock. I want, however, to point out that although the cost of pressed steel, the "tubular" or any other system for effecting large reduction in tare weight, is undoubtedly greater per ton weight of the vehicle it is actually less per ton weight of the load. The design and carrying capacity of a vehicle is the proper basis on which to compare the relative cost of wagons. In purchasing wagons of a heavy tare you actually pay more for what you want to get rid of. I am, of course, comparing the cost of pressed steel with wagons of ordinary construction, which have been designed and built to give good results in working and maintenance, and not with some light railway stock with which I am acquainted, which is built only to sell.

The wagon stock for all classes of work is of the bogie type, 25ft. long over headstocks, and 7ft. wide. The length over buffers of each vehicle is 28ft. 3in. The centres of the bogies are 16ft. 8in. and their wheelbase 4ft. 3in. The weight complete on rails of each wagon fully loaded is 20 tons or 5 tons per axle. The bogies of both carriage and wagon stock are of the swinging bolster type and are identical and interchangeable. A few wagons have been fitted with bogies without the swinging bolster, for experimental purposes, but these bogies are interchangeable with the remainder.

Scene inside Leeds Forge Works. Underframe of steel wagon for 2ft. 6in. gauge. 40ft. long 7ft. wide. Tare of completed wagon 6 tons. Designed to carry 14 tons.

The Timmis system of springs has been adopted throughout the carriage and wagon stock with much success in obtaining smooth running. The system consists in the use of duplex spiral springs, so arranged that when the vehicle is running empty, one spring only of each set is in action, but when running loaded both are called into play. This is a point of considerable importance for very light stock, since in the absence of this arrangement a wagon when unloaded would run virtually without springs, and so be subjected to unnecessary vibration, which means useless wear and tear.

Low Side Wagon loaded with sacks.

The above illustration shows a low side wagon loaded with sacks containing sand. The first layer of sacks are placed with their ends lying upon the low side. This gives them an upward tilt which is communicated through each of the succeeding layers up to the top, so that bags and sacks loaded in this way cannot fall off. This wagon has sufficient capacity to carry full loads (15 tons 18 cwt.) of wheat, seeds, and pressed cotton, and is the best for general utility. The low side facilitates loading and unloading. It will be seen that the ends of this wagon are stamped from one plate. The brake acts on both wheels of one bogie and can be operated from either side of the wagon. This wagon weighs 4 tons 2 cwt. and carries 15 tons 18 cwt.

The Barsi Rolling Stock

The next illustration shows a high side wagon loaded with coal. It is fitted both with side and end doors, the latter to enable guns and limbers to be carried. Owing to the end doors and their brackets, a wheel and chain is substituted for the end brake lever. This gear can also be operated from

Above: High side wagon loaded with coal.

Below: All steel covered wagon.

either side. The sides of these wagons are stamped and dished in sections. The dishing should be sufficient to deflect a bullet, while the flanges prevent splashing. This wagon weighs 5 tons 7 cwt. and carries 14 tons 13 cwt.

The covered wagon, built entirely of steel, weighs 5 tons 18 cwt. and carries a load of 14 tons 2 cwt. Its capacity is 1,000 cubic feet. The body is 7ft. wide and 6ft. 6in. high inside, and is designed to carry six horses with forage and attendants, and at a pinch it can carry nine.

The Barsi Rolling Stock – Carriages

The carriage stock for all classes of work is also of one length, namely, 40ft. 6in. over bodies, and 40ft. over headstocks. The bodies are 7ft. 6in. wide over sunshades and 6ft. 2in. between standing pillars. The underframes for every class of vehicle are identical in every respect and interchangeable. The bogies are 28ft. centres, with a wheelbase of 4ft. 3in. and, as stated, are interchangeable with those under the wagons. The cars for ordinary use are built to weigh, with a double complement of passengers and baggage, rather less than 20 tons on rail, which is equal to 5 tons per axle. They are capable of taking curves of 150ft. radius. The cars are supplied with gas fittings throughout. There are only two classes, namely, upper and lower class, but the lower class compartments are of a style and finish equal to that usual in the second class. We believe in catering for the comfort of our lower, or third class passengers, from whom we shall derive a large revenue.

A passenger train made up of one upper class car, eleven lower class cars, and one compound brake van, allows full seat room for 30 upper class, and 736 lower class passengers. In times of pilgrimages, or on other special occasions, one thousand passengers could be carried in a single train.

The Special Saloon Car contains a main saloon, a smoking or inspection compartment at the end, an entrance and luggage vestibule, a lavatory with a bath and wash basin supplied with hot and cold water, a kitchen and servants' compartment. The latter communicate with the entrance vestibule by means of a side corridor. The Car has sleeping accommodation for six passengers. These and all other cars have sunshades and double roofs, gas lights, louvre shutters, and venetian ventilators. This Car weighs 12 tons 13 cwt. all complete.

The Compound Brake Vans have been built with upper and lower class compartments to enable passengers to be carried by every goods train. The upper class compartment has seats for six passengers and sleeping berths for four, and is furnished with lavatory accommodation. The lower class compartments are fitted with cross seats and carry 32 passengers. The Brake Vans weigh 12 tons 1 cwt. each. In the lower class compartments the cross seats carry four passengers each comfortably, and at a pinch five.

The lower class cars have lower class compartments only, and carry 64 passengers. They weigh 11 tons 13 cwt.

Barsi Light Railway - above: Special Saloon Car.

Below: Brake Van.

Means for avoiding Transshipment at Junctions between Broad and Narrow Gauge Railways

The value of narrow gauge in reducing cost is so great that any practical proposal, which promises to avoid the cost, loss of time, and damages to goods caused by transshipment, and yet permit the use of narrow gauge for short branch lines, will, in view of the light railway projects now being matured in England, at all events be regarded with interest. I propose to effect this by means of Transportation Cars, and Transfer Bogies; the former being intended for general use in England, the latter to meet special conditions in India and abroad.

The Transfer Bogies are placed under standard gauge wagons after their own wheels have been removed. The bolster of the narrow gauge bogie is furnished at each end with a dummy axle-box, sliding in the axle-guard of the standard gauge wagon, which is therefore carried on its own springs. The Transfer Bogies, however, are not suitable for light railways in England where the average length of line is so short, but have been devised for a special purpose, namely, with the object of carrying standard gauge wagons loaded with coal, which suffers much in transshipment, over long lengths of metre gauge line amounting in the case of the Southern Mahratta Railway to hundreds of miles. These coal wagons would come back to the junction and to there own wheels and axle-boxes, and the short time occupied in the exchange, effected by means of an hydraulic lifting plant, would be justifiable and economical.

An experimental four-wheeled Transportation Car, tested at the Newlay Exhibition, was constructed to demonstrate the feasibility of carrying by this means loaded lorries from Liverpool to Manchester on a narrow gauge line, so as to avoid all transshipment between ships' side at the Liverpool Docks, and the warehouse in Manchester and *vice versa*. This project is now under the consideration of a special committee of the Liverpool Chamber of Commerce, appointed to investigate means for transporting traffic in competition with unduly heavy rail charges. The result was a successful demonstration of the stability of the cars, and of the feasibility of the project as far as these mechanical proposals were concerned.

The Barsi Rolling Stock as Applicable to Military Operations

The theoretical requirements of a field railway, which I state with the diffidence which becomes a mere civilian, appear to me to be that generally the equipment must include permanent way, portable bridges, rolling stock, and all stores of whatever kind that may be required in any part of the world for use on an unsurveyed route through a country unprovided with railways, and so as to be entirely independent of local resources, and, in particular, as follows :-

1. The permanent way to be portable and of a pattern which takes the fewest men to handle and the shortest time to put together, and to be of the least weight and bulk compatible with the largest traffic capacity.

2. The rolling stock to have a uniform maximum axle-load, and to have the greatest load capacity with the lightest tare weight, compatible with a proper reserve of strength.

3. The wagon stock while adapted to the greatest number of purposes and contingencies to be of the fewest types. Details, as far as possible, to be common to and interchangeable in all types, and so that repairs in the field can be effected by the exchange of standard parts.

4. The rolling stock to be of such patterns that their erection on arrival should be an expeditious operation requiring neither tools nor workshops.

I wish to point out briefly to what extent the Barsi rolling stock fulfils these conditions.

As regards No. 1, both rails and sleepers stow perfectly for shipment. A 24ft. length of permanent way including ten steel sleepers can be put together *in situ*, keyed and fished by six men in about five minutes, when working systematically. The permanent way can also be carried in the wagons, put together with sleepers complete in 24ft. lengths ready for laying down. The weight of permanent way, with 30lb. rails and 40lb. sleepers, with the usual additions for contingencies and waste is 90 tons 16 cwt. per mile. The net load of stores which can be taken on a gradient of 1 in 100 is 210 tons, so that with 13 low side wagons and one brake a single train can carry two miles of track per trip, with a full complement of workmen. The Barsi rolling stock can transport in either direction a maximum of 3,360 tons daily over a section 20 miles long, with a ruling gradient of 1 in 100, and a maximum of 1,680 tons with gradients of 1 in 50 on curves of 250ft. radius. This is on a basis of 16 trains a day in either direction, which, with crossings, can be worked without difficulty. As regards troops, a battalion with stores and impedimenta could be carried in one train over grades of 1 in 100, and a half battalion over grades of 1 in 50.

As regards No. 2, that the rolling stock shall have a uniform maximum axle-load, and the greatest load capacity with the lightest tare weight compatible with a proper reserve of strength, the Barsi rolling stock represents, I believe, an advance on anything which has been done hitherto in these directions as regards light railways. The wagons are so light that they reduce the weight and bulk of permanent way to the smallest proportions, thereby increasing the length of track which can be carried in one ship, or by one train, and the permanent way being lighter, it is more easily handled, and laid quicker.

As regards No. 3, the Barsi wagon stock has been restricted to three types, which are all that are necessary for ordinary traffic purposes. The parts which are subject to wear to tear are identical and interchangeable throughout.

With reference to No. 4, that the rolling stock should be of such patterns that their erection on arrival should be an expeditious operation requiring neither tools nor workshops, the whole of the rolling stock has been erected in India under these conditions. The pressed steel underframes require no trusses, and very little packing, consequently they stow closely on board ship, and the risk of breakdown from lost truss-bolts and nuts is avoided. Their lightness enables them to be shipped in large pieces, and to be handled with very elementary lifting appliances, so that the time lost on arrival, in erecting stock shipped in small sections for the greater convenience of handling, is avoided.

As regards landing carriage and wagon stock from vessels, the underframes can be lowered forthwith on to the bogies, which can be shipped complete with wheels, direct from the ship's side, and the cases containing sides and ends can then be lowered on to these underframes as they stand. The underframes as soon as they are on their bogies already form a movable wagon, and can be rolled away to sidings where their erection will be completed. One engine could be carried erected complete on deck, and slung down on to rails or run down an inclined plane, so as to be ready for the first lot of wagons when put together. It is of the greatest importance to keep the landing clear of accumulations of stores.

The Barsi carriage stock has been specially designed to allow of speedy erection after shipment. The floors, sides, ends, roofs, and partitions are shipped in single pieces. The partitions are packed in a square case by themselves, and the two sides complete, with doors and sunshades, in one long case by themselves. With extemporised lifting appliances only these cars were put together in running order on rails within three days from the time of unpacking the cases. Military carriage stock would be of the same length but of a simpler character, and could be put together in a few hours.

The object of this system is to leave the parts as complete as possible so that the work of erection on arrival is made as simple and expeditious as possible. For expeditious work I do not believe in the system of designing rolling stock for military purposes out of permanent way material, with small units which can be converted into larger units and so on. The whole thing becomes a large Chinese puzzle, and unless you have men who are thoroughly acquainted with all the parts and their many combinations the result is too often a big muddle. What is wanted is the fewest number of parts, and extreme simplicity throughout so that it is perfectly obvious to an ordinary mechanic where everything has to go without requiring either drawings or instructions.

To be of the fullest use a military railway must keep abreast of the advance of the troops, which means progress at an average rate of ten miles a day. I regard this as a practicable aim provided the work of construction proceeds with a maximum speed and efficiency from the first day and hour of debarkation. Twenty miles lost, by avoidable delay at the start, may never be made up. It is only continued practice that can ensure the smart start off. No railway corps can make a proper show on service unless all the operations of disembarkment, erection, alignment and construction are as familiar as are the ordinary evolutions on the parade ground. Once this familiarity is attained under conditions approximating to those existing on active service I see no reason why a light railway could not be laid down at a speed much greater than that which has as yet been attained in the field.

Addendum

The Barsi 2ft. 6in. gauge Light Railway was opened for traffic in March last, and the above paper was read about a month previously. Sufficient time has therefore hardly yet elapsed to afford a thoroughly conclusive test of the value of the principles upon which both permanent way and rolling stock have been designed; but the experience of the working of the railway during a period of an almost unexampled combination of troubles, including plague, cholera, and famine, and their ruinous effects upon traffic has been sufficient to indicate that the importance claimed for them, from the commercial point of view, has been overrated.

It is a satisfaction to be able to state, notwithstanding these troubles, that the working expenses in India were no greater than 49.99 per cent. of the gross receipts, and that the net profits earned were about 4 per cent. on the capital of the Company. Under normal conditions of traffic, and more particularly when the extensions of the lines are completed, I have no doubt that the percentage of working expenses will be reduced. The Company have the good fortune to possess in Mr. A. L. Alexander, M.I.C.E., a resident engineer and agent distinguished alike for his capacity, tact, and energy. His most efficient management claims grateful acknowledgment and must be taken into consideration with any suggestion that the power of the engines and the light tare weight of this rolling stock have been the chief factor in keeping down working expenses to the above-mentioned low figure.

As regards the bulk capacity of the Barsi Rolling Stock in practice, the resident engineer and agent reports in a letter just received that he has carried the whole of the machinery and buildings of a ginning factory in one special train*. Amongst the machinery was a boiler 28ft. 6in. long and 7ft. 6in. in diameter and weighing 14 tons, carried on one wagon, and a fly wheel 10ft. in diameter and weighing 4 tons carried in another. He concludes by saying:- "It is worthy of note, that this is the first time in India that such a weighty piece of machinery as the boiler has been carried on a 2ft. 6in. gauge railway; also that (I believe) ours is the only railway in India, broad or

---

*See photograph of this train on page 1.

narrow gauge, where one piece of machinery of the dimensions previously quoted has been carried on an ordinary goods truck." On the arrival of the boiler at the junction with the Great Indian Peninsula Railway, which is of a gauge of 5ft. 6in., it had to be placed on two broad gauge wagons specially sent up from Bombay to receive it.

With reference to the bulk and weight capacity of the wagons for ordinary traffic, such as wheat, linseed, salt, cotton, piece goods, etc., the following table gives the results at present attained, which, as regards open wagons, are being gradually improved by better stowage :—

Capacity of Barsi Wagon Stock in Practice

Type of Wagon	Tare Weight T. - C.	Maximum Load T. - C.	General Traffic Average T. - C.	Linseed in Bags No. of Bags	T. - C.	Fully pressed Cotton Bales No. of Bales	T. - C.
Covered	5 - 18	14 - 2	14 - 2	156	14 - 0	78	14 - 2
High-Side	5 - 7	14 - 13	14 - 10	157	14 - 2	74	13 - 15
Low-Side	4 - 2	15 - 18	14 - 18	160	14 - 7	74	13 - 15

The Secretary of State for India has accorded sanction for the extension of the railway to Pandharpur, a further distance of 31 miles, and included in the rolling stock required will be two transportation cars for carrying 5ft. 6in. gauge wagons without transferring their contents. These have also been designed particularly with reference to military requirements, for carrying guns, howitzers and commissariat vehicles. As the wheels of the carried vehicles will be only nine inches above rail level, they will be capable, on occasion, of very rapid entrainment and detrainment.

Transportation cars of a somewhat similar character will be employed on the Welshpool-Llanfair 2ft. 6in. gauge Light Railway, recently sanctioned by the Light Railway Commissioners. In this case they will be used for the transportation of fully loaded 4ft. 8½in. gauge railway wagons over the narrow gauge line. The advantages of avoiding transshipment in England are considerably greater than in India, where the low cost of labour makes it generally preferable to transship goods rather than incur the haulage of additional dead weight.

Ten wagons of the Barsi low-side type has been ordered by the War Office for experimental purposes, and will shortly be completed. Following on the results already achieved by the Barsi Light Railway, the *Times of India* and other Indian newspapers are now strongly advocating the immediate construction of light railways of the Barsi type as a means of holding and pacifying the North-West Frontier.

With reference to the system of standard details, by which all the parts of rolling stock are coded and numbered, a case occurred, shortly after the opening of the line, of a derailment of a locomotive, through a station-master omitting to lock the points; and as it was travelling at some speed with a heavy train behind it considerable damage was done to the engine. The numbers of the damaged parts were at once cabled home, and the order for their renewal was in the hands of the makers within three days, and, but for the engineers' strike, the renewed parts would have been on their way to India within two or three weeks of the accident.

E. R. Calthrop